Bibliographic information published by the German National Library:

The German National Library lists this publication in the National Bibliography;
detailed bibliographic data are available on the Internet at http://dnb.dnb.de .

Imprint:

Copyright © 2015 GRIN Verlag
Print and binding: Books on Demand GmbH, Norderstedt Germany
ISBN: 9783668100534

This book at GRIN:

https://www.grin.com/document/311402

Anonym

The 10 Most Habitable Planets

GRIN Verlag

GRIN - Your knowledge has value

Since its foundation in 1998, GRIN has specialized in publishing academic texts by students, college teachers and other academics as e-book and printed book. The website www.grin.com is an ideal platform for presenting term papers, final papers, scientific essays, dissertations and specialist books.

Visit us on the internet:

http://www.grin.com/

http://www.facebook.com/grincom

http://www.twitter.com/grin_com

10 Places that might have alien life

Content

Intro:

Finding life outside of Earth has played an enormous part in the history of Space. So far we have no proof that life exists beyond our own planet, although, after a lot of work, scientists have surprisingly found places that could, in fact, sustain life. This, of course, isn't guaranteed, but, if we were to live there, we would do just fine.

In the following list, I will be writing about the top 10 places that could quite possibly have alien life. Some of these places can be as far 1,000 light years or as close as Jupiter, keep reading to learn more about these extraordinary places.

List:

10. Tau Ceti F

Tau Ceti F is one of the two planets located at it's solar system's habitable zone. The planet lies just 11.9 light years away from us, making it one of Earth's neighbor's. The planet is believed to have water, and it is about 4.3 times bigger then the Earth, causing it to have more gravity, thus, forcing it to keep most of it's water, right?

Well yes, only if the planet was a couple billion years older. Study from NASA's telescopes, suggests that the planet has only recently moved to the goldilocks zone, making it an inhabitable planet for a couple more billion years.

9. HD 85512 b

This weirdly named planet orbits it's star, HD 85512, around 35 light years away from Earth. The recently discovered planet is located at hottest part of the Goldilocks zone, and it is believe that liquid water can be found on it's surface.

In order for this fascinating planet to be habitable, for humans, or for alien life, it has to contain at least 50% of cloud (water vapor). So far, the planets atmospheric composition is still unknown, therefore, it is very unlikely for life to form on this planet, but, compared to most of the exoplanets found, it's still one of the most habitable planets.

8. Gliese 581 D

This planet Is just 20 light years away from Earth, which is the distance light travels in 20 years, and it is currently orbiting the Gliese 581 red star, which is famous for the discovery of some likely habitable planets. The star is located at the outer layer of it's solar systems goldilocks zone (Habitable zone) thus, causing it's temperature to drop to an average of -18C. This is also caused by its lack of atmosphere, which here on Earth, helps heat up our planet.

Gliese 581 D isn't expected to have advanced alien life, although, it is said that the lowest temperature life can withstand is -20 degrees Celcius, therefore, there is a small chance that microscopic extremophiles (organisms that can survive in extreme enviroments) are living there.

7. Gliese 667 Cf

There is very little known about this cosmically insane planet, one of the craziest things about this planet, is that it's part of a rare three sun system. This super earth is currently orbiting at the goldilocks zone of its red dwarf, which is approzimately 31% the size of our own sun. Also, the planet is orbiting 18 million kilometers away from it's red star, therefore, it is orbiting much closer to it's sun then Earth is. The planet is slightly further then 22 light years and due to the fact that it's a super Earth, there is an enormously high chance of it being a rocky world.

The temperature is expected to be 30C in average, which makes it quite likely to have water, although, the planet, has to have at least 50% water vapor in order to sustain life, either wise, it will be as dry and sandy as Mars. It is believed that this planet has a similar atmosphere to Earth, so therefore, it probably has that 50% of water vapour.

6. Kepler-62e

Again, this Earth-like planet, is located at the goldilocks zone and is about 1.6 times bigger then our own planet. One year, is about 122 days, meaning that you would have to buy 3 times more calendars if you lived there. This planet is believed to be a rocky world, therefore, it is unsurprisingly likely for it to have water.

Also, the planet probably contains oxygen and it is always warm and humid, except in it's ice poles. It is very likely that advanced life has formed there.

5. KOI 736.01

KOI stands for kepler object of interest due to the fact that this is one of the most likely habitable planets discovered. This fascinating planet is just 0.12G more then Earth, and is also 1.6 times bigger. There is very little known about this planet, but it is believed that it's similar to Mars and that it has water, weather and oxygen.

The planet is considered one of the most habitable planets based on the habitability rating system. The planet is a meso-planet meaning that it's neither too big nor too small. It has a rocky surface and it's temperature is 0 degrees celcius minimum and 50C maximum. As I said earlier, has a similar atmosphere with Earth and it is expected to have oxygen. Like all of the planets in this list, it is located in the goldilocks zone.

4. Kepler 69C

This is one of the most fascinating planets we have found. It is believed that this planet is orbiting at the most habitable part of the goldilocks zone. This Venus-like planet has a high chance of containing essential gases such as oxygen, as well as the most essential liquid of all, water.

In space, it is easy to find water, although, it is hard to keep it in it's liquid state. Fortunately, Kepler 69C is about 70% larger then Earth meaning that it has more gravity. This extra gravity, has helped Kepler 69C keep it's water, thus, making it an ocean-world. And as far as we know, water = life

Sadly, this planet is an astonishing 2,700 light years away from Earth, meaning that reaching it with the current technology will take tens of thousands of years.

3. Kepler 186f

Seems to me that the Kepler telescope has found a lot of habitable planets...

Kepler 186f was one of the first Earth-like planets to be discovered, and it is still considered one of the top planets that can support life. There is a very high chance of the planet being a rocky world, and it is located at the best spot of the goldilocks zone. Due to it's Earth like size, the planet is likely to have a similar gravity, therefore, it is highly likely that it is partially covered with oceans, just like the Earth.

The planet is located at the constellation Cygnus and it is orbiting the star Kepler 186, which is approximately 500 light years away from our own planet. Reaching this world would be a huge milestone, due to it's enormous distance but also because of it's huge chance of habitability.

2. Kepler-22b

Kepler 22b has a radius of about 2.4 times bigger then Earth. According to observations from infrared telescopes, this planet is highly likely to be a water world, just like Kepler 69C. The reason I ranked this planet higher, is due to the fact that a lot more is known about it. This earth like planet's temperature is -11C at worst cases, and 22C at best cases, meaning that it's friendly to advanced alien life. Also, the planet is believed to experience a tiny greenhouse effect, which is probably why the temperature is so life friendly.

If we have to move to another planet, Kepler 22b is the right place and if we do, we will hopefully find tons of fish to eat. Unfortunately, going there will be extremely expensive and will require hundreds of tons of fuel as well as food.

Kepler 22b is one of the most habitable planets ever discovered, although, just like all of the places in this list, habitability isn't guaranteed and we will never know until we go there.

1. Europa

If you already don't know, Europa is Jupiter's icy moon and it is located between Io and Ganymede. According to research done by various NASA probes, Europa is believed to have a whole ocean under it's thick icy crust. Below this icy crust, thousands of fish could survive and live, due to the fascinating fact that Europa has more then 40% oxygen in it's ocean.

Luckily, NASA is said to be launching a probe to orbit the moon, if the mission is successful, NASA will later on send a small rover with a tiny submarine that will hopefully drill down to the ocean.

NASA has already selected nine scientific instruments for the first mission, the selected instruments are:

- Plasma Instrument for Magnetic Sounding (PIMS)
- Interior Characterization of Europa using MAGnetometry (ICEMAG)
- Mapping Imaging Spectrometer for Europa (MISE)
- Europa Imaging System (EIS)
- Radar for Europa Assessment and Sounding: Ocean to Near-surface (REASON)
- Europa THermal Emission Imaging System (E-THEMIS)
- MAss SPectrometer for Planetary EXploration/Europa (MASPEX)
- Ultraviolet Spectrograph/Europa (UVS)
- SUrface Dust Mass Analyzer (SUDA)

Sources:

Link 1: http://listogre.com/facts/science/top-10-earth-like-planets/

Link 2: http://www.space.com/12918-habitable-alien-planet-hd-85512b-super-earth-infographic.html

Link 3: http://www.planetsedu.com/habitable-planet/hd-85512-b/

Link 4: http://www.space.com/29191-exoplanets-tau-ceti-alien-life.html

Link 5: http://astrobiology.com/2015/03/chemical-composition-of-tau-ceti-and-possible-effects-on-terrestrial-planets.html

Link 6: http://www.solarsystemquick.com/universe/gliese-667cc.htm

Link 7: https://www.nasa.gov/mission_pages/kepler/multimedia/images/kepler-69-diagram.html

Link 8: http://www.nasa.gov/mission_pages/kepler/news/kepler-62-kepler-69.html

Link 9: http://andrewrushby.com/2011/12/05/the-habitable-exoplanets-catalog/

Link 10: http://www.jpl.nasa.gov/missions/europa-mission/

Link 11: http://www.lunarplanner.com/Images/planets/Jupiter-Main-Moons.jpg

Link 12: http://news.discovery.com/space/alien-life-exoplanets/most-earth-like-alien-world-discovered-140417.htm

Link 13: http://www.space.com/8487-jupiter-moon-ice-covered-ocean-rich-oxygen.html